Our Attribute Walk

by Luana K. Mitten and Mary M. Wagner

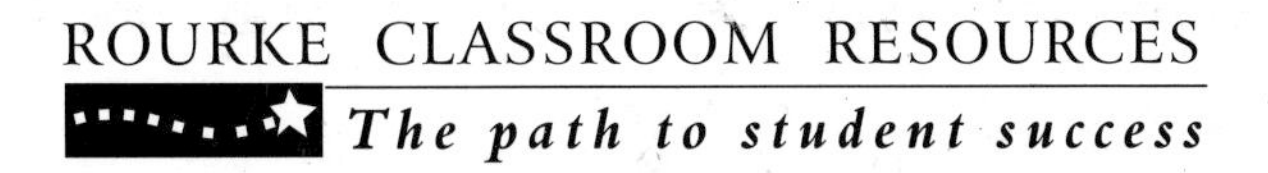

Big and little, old and young, heavy and light, ... **attributes** help us describe everything around us.

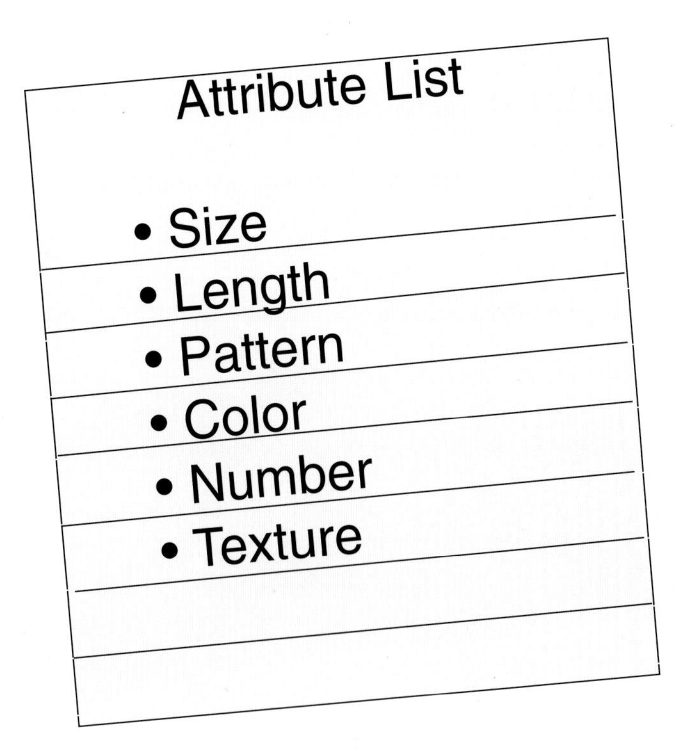

Let's take an attribute list with us as we look at the world of nature.

Look! Do you see the huge whale? Whales are the largest **mammals** in the world.

Can you see the little squirrel? A squirrel is a small mammal.

Do you see the giraffe with a long neck? Its long neck helps it eat leaves from tall trees.

Do you see the prairie dog with short legs? Short legs help the prairie dog tunnel in the dirt.

Do you see the striped zebra? The pattern helps protect it from **predators**.

Do you see the spotted jaguar?
The jaguar's pattern helps it hide while it hunts.

Do you see the yellow flowers?

Flowers can be many different colors.

Can you count three ants and four birds?

Do you see the fluffy chicks and smooth eggs?

We found many attributes on our **make-believe** walk. What attributes can you find on a nature walk around your school?